Mathematical Principles of l

Edition 1

M. K. Jones

* * *

(blank page)

Mathematical Principles of Psychology
Edition 1

M. K. Jones

Introduction

By '**mathematical principles of psychology**' we don't mean anything along the lines of the correlation coefficients and such of Experimental Psychology, but something much more along the lines of what **Kurt Lewin** was doing in his '**Principles of Topological Psychology**' (published in 1936). (- a tip of the hat to **Frank Harary**, who included mention of Lewin's book in the bibliography of his textbook '**GRAPH THEORY**' – which is how I first became aware of Lewin's work.)

As **Edna Heidbreder** notes, in her '**Seven Psychologies**' (published in 1933), man's scientific attention has progressed from far outwards (Astronomy) inwards, towards Physics, then Chemistry, then Biology, and finally, to Psychology. Each successive inner level subsumes, or at least presumes, all knowledge of the previous level. It follows, therefore, that a comprehensive knowledge of Psychology would more or less amount to a **Theory of Everything** (Lewin's disparaging remark (p. 21) about the **philosophical Utopia** of a **universal science** notwithstanding – Lewin, although right on the money most of the time, is simply picking up the wrong end of the stick in this case.).

Actually, there is one more discipline that should be added to the progression: **Linguistics**, situated between Biology and Psychology. Linguistics is so bound up with Psychology that many aspects of Psychology can be illustrated within Linguistics, much like, in mathematics, groups can be studied by studying their representations as matrices. (The psychological experiments regarding memory, mentioned by Heidbreder, strongly point in this direction.)

also:
"even introspection illustrates the tendency of observation to work from without inward"
-- Edna Heidbreder, Seven Psychologies (p. 101)

There are actually **two currents** of inquiry involved, the superficial, and the deep, and they run in opposite directions. Scientific psychology is the deep current. The superficial current is described by Winwood Reade in his book 'The Martyrdom of Man':

"It is the nature of man to reason from himself outwards."
-- Winwood Reade, The Martyrdom of Man, p. 84

Notice also that physical phenomena progress from local uniqueness (e.g., a gas at low temperature or high pressure) to global uniformity (e.g., a gas at high temperature and low pressure). Man is called upon (by God, or by the nature of things) to progress in the opposite manner, that is, from local uniformity to global uniqueness ("post longa laboro, atingos la celon en gloro" – L. L. Zamenhof), and to the extent that he fails in this he is corrupt / sinful, and such corruption / sinfulness eventually leads to war, the efficient cause of war being greed, but the final purpose of war being to change global uniformity into local uniformity (e.g., curfews) and local uniqueness into global uniqueness (e.g., industrious farm boys becoming famous generals).

"If we promoted justice and charity among men, we should be playing directly into the Enemy's hands; but if we guide them to the opposite behaviour, this sooner or later produces (for He permits it to produce) a war or a revolution, and the undisguisable issue of cowardice or courage awakes thousands of men from moral stupor. This, indeed, is probably one of the Enemy's motives for creating a dangerous world—a world in which moral issues really come to the point. He sees as well as you do that courage is not simply one of the virtues, but the form of every virtue at the testing point, which means, at the point of highest reality. A chastity or honesty, or mercy, which yields to danger will be chaste or honest or merciful only on conditions."
— C.S. Lewis, The Screwtape Letters

Uniqueness is always immediately profitable (and hence to temptation to engage in it upfront, especially by way of being a spectator), but uniformity may or may not be immediately profitable. A major part of the uniformity that is not immediately profitable is study. Persisting in uniformity that is not immediately profitable is called faithfulness. ("O come all ye faithful, joyful and triumphant...") One cannot obtain an income without posting to foreground, and study greatly facilitates such posting. Hence, virtue is its own reward.

Blaise Pascal, when he said, "All the trouble in the world is due to the inability of a man to sit quietly in a room." was advocating submission to local uniformity. And Bob Greene's admonition, "Never travel for food." is specific injunction towards local uniformity. And on p. 319 Heidbreder says, "without absorption...there is no first-rate accomplishment". It easy to take "absorption" to be "local uniformity" and "first-rate accomplishment" to be global uniqueness. And there is poster that says, "Eat. Sleep. Read." The admonition, "Publish or perish." also comes into play here, publication being the modus operandi of global uniqueness.

The doctrine of original sin is reference to the fact that a baby is shrouded in uniqueness, which, being local, is, of course, is a violation of the precept of local uniformity. One can then play off this for supporting a claim to divinity, as is done in Christianity. Jesus was free from original sin because the uniqueness he was shrouded in was not local, but global, and his mother was free from original sin (the 'immaculate conception') because she was to be the mother of God, and so her destiny, from the very onset of her life, was global. (The reality is, of course, that Jesus never existed. The New Testament is a fabrication from start to finish. It was written, in secret, by the Roman aristocracy as an antidote to Judaism, Jewish insurrections being a thorn in the side of the Roman empire. When Martin Luther advocated severe sanctions against the Jews, he was not being 'unchristian', but was acting consistently with the original intent of Christianity: to undermine Judaism. Consistent with the fact that Jesus of Nazareth never existed is the fact that there is no prohibition against depicting his likeness. The situation is different for real persons. For example, in China it is illegal to use the depiction of Mao Zedong for marketing purposes. And regarding the Old Testament, one may ask why the scribblings of drunken goatherds should be taken as Holy Writ.)

A: "Have you read the Bible?"

B: "Yes, and I've also read Aesop's Fables."

Man has always had an innate intuition that he should be uniform locally, in order to obtain uniqueness globally. This explains the enduring popularity of religion. Religion, with its promise of paradise (global uniqueness) greatly aids its adherents in achieving local uniformity via its rites. Of course, as Goethe noted, those sufficiently culturally mature have no need of religion.

"He who possess art and science has religion; he who does not possess them, needs religion."
-- Goethe

We accept as our starting point, as did many eminent thinkers of the past (as Heidbreder points out), the world at it naively appears to our intuition. We will immediately add to that, however, knowledge of mathematics.

"As a natural science, psychology must begin at the starting point of all science, the world as it appears to naive, everyday observation."
-- Edna Heidbreder, Seven Psychologies (p. 175)

We can define mathematics as any dealing with **distance**. If the dealing is **internal**, we default to getting the familiar sequence of topics of Arithmetic, Algebra, and Calculus. If the dealing is **external** (at arm's length, so to speak), then we default to getting things like General Topology and Abstract Algebra. However, we sometimes wish to override the defaults. If we deal with internal distance at arm's length, we get what we can call 'quantitative qualitative' models, that is, qualitative models that use quantitative terminology, and some of what we will be doing herein will be that (specifically, what I call the 'partition' function). If we deal up-close-and-personal with external things, we get identification of the mathematical structure with the corresponding real-world reality. Kurt Lewin explicitly promoted this approach:

"I remember the moment when – more than ten years ago – it occurred to me that the figures on the blackboard which were to illustrate some problems for a group in psychology might after all be not merely illustrations but representations of of real concepts."
-- Kurt Lewin, Principles of Topological Psychology (p. vii)

also:
"I am convinced that these concepts which we use for the representation of psychological facts, like region, spacial relationship in life space, connectedness and separateness, belongingness, etc., are real spacial concepts in a strict mathematical sense."
-- Kurt Lewin, Principles of Topological Psychology (p.42)

also:
"The nature of the things whose system constitutes a mathematical space is entirely irrelevant for modern mathematics. ... Only certain relationships and the possibility of certain operations are relevant. ... As far as mathematics is concerned there is therefore no fundamental objection to applying the fundamental concept of space to psychological facts."
-- Kurt Lewin, Principles of Topological Psychology (p.52)

Before we take up the mathematical approach, it is worth mentioning the well-known phenomenon of the automatic emergence of structure out of randomness. There is even a simple example involving psychology, namely, that at a party, among any six people chosen, there will be at least three that are mutual acquaintances, or three that are mutual strangers. There is an entire theory devoted to such phenomena: **Ramsey theory**. We won't be pursuing it herein, but the point is that we do not always have to impose order, but sometimes merely describe order, perhaps hidden, that already exists. Indeed, Lewin as much as alludes to this phenomenon:

"This supplies the empirical premise for the application of the topological concepts to the life space." (i.e., because of structuring that certainly occurs)
-- Kurt Lewin, Principles of Topological Psychology (p. 62)

Chapter 1: First considerations of mathematics

Chapter 1, Section 1: Positive measure

As our first insight from bringing mathematics to bear on the world we note that physical quantities, such as age and volume, always have positive measure, and therefore the appropriate way of comparing them, as compare we must, is by division, not by subtraction. For example, how does a 10-year-old compare with a 1-year-old and a 100-year-old? Is the 10-year-old closer in age to the 1-year-old? No, the 10-year-old is equidistant between a 1-year-old and a 100-year-old, because the 10-year-old is 10 times older than the 1-year-old, and the 100-year-old is 10 times older than the 10-year-old. Saying that a 10-year-old is 9 years older than a 1-year-old is literally true, but technically false. (The same goes for saying that your salary is such-and-such amount of dollars per year. Actually, your salary is a certain FRACTION of the total wealth of the society that you are in. Since the literal truth is easier to handle linguistically, the literally true is what we accept, and expect, in ordinary language, just like using the terms 'sunrise' and 'sunset' even though we know those terms are not literally correct.) So, the regime of measurement is 'multiplicative' instead of 'additive'. (The term "geometric" is normally used for multiplicative regimes in this sense, because the main geometric parameters of interest – namely, area and volume – are obtained by multiplication, the term "multiplicative" being reserved for certain types of functions within Number Theory.) However, an additive regime is much easier to deal with. However, knowledge of mathematics extricates us from this impasse, because we know that the multiplicative regime of the positive real numbers is **isomorphic** to the additive regime of the arbitrary real numbers. The isomorphism is called 'the logarithm' (mathematical notation: 'log'), when going from the multiplicative regime to the additive regime, or 'the exponential' (mathematical notation: 'exp'), when going from the additive regime to the multiplicative regime. (The exponential, because its role is so often to return to the original venue after applying the logarithm, is also called 'the antilog'.) Since the logarithm and exponential are inverses of one another, each one composed with the other is the identity function. Denoting the identity function by 'i' and function composition by '..', we therefore have: exp..log = i, and log..exp = i. Also, we will employ the convention of ordinary language in using the definite article ('the') for things that are, strictly speaking, not unique. Logarithms (and exponentials) vary depending on their base, but we still say "the logarithm", just as we say "They took him to **the** hospital.", when in fact there are several area hospitals. We further employ the convention that the logarithm in question, unless stated otherwise, is the **natural logarithm**, and that when a logarithm and an exponential are employed in tandem, as they so often are, that they have the same base. Remember that the exponential is often present only implicitly, as in the expression 'take logarithms' as the advice for solving a problem: 'take logarithms' is short for 'take the logarithm, do the intermediate calculations, and then take the exponential'. For example, we can compute xy by the sequence xy = exp(log(xy)) = exp (log(x) + log(y)).

Chapter 1, Section 2: Tug-of-war between extremes – resolved by mathematics

The phenomena encountered and described by psychology and elsewhere is often a tug-of-war between extremes. Heidbreder describes one such situation on page 81:

"He noted that sensation and stimulus apparently do not increase in intensity at the same absolute rate. For example, if one candle is burning in a room and another candle is lighted the difference is immediately noticed; but if ten candles are burning and one more is added, the difference is scarcely appreciated. It is possible, then, that sensation may increase in arithmetical progression as the stimulus increases in geometric proportion."
-- Edna Heidbreder, Seven Psychologies (p. 81)

The situation is one of starting out at near 100% response and falling off asymptotically to 0% response at infinity. However, this is how the logarithm behaves in the range from 1 to infinity. The logarithm rises to infinity as its argument rises to infinity, but exceedingly slowly (specifically, more slowly than any polynomial), and, since the logarithm closely matches the identity function at 1, provides a model for, among other things, quantity pricing, and for how partitioning depends on measure. All we have to do is shift the logarithm one unit to the left to get this model. This is done by adding 1 to the argument. So, what we are dealing with is $\log(1 + x)$. Let us give this expression a concise name. We will call it $r(x)$, and we will call its complement, namely, $x - r(x)$, $r'(x)$. Then, $r(x)$ gives quantity pricing for when an amount x is purchased (with $r(x)/x$ giving the decreasing unit price), and the complementary ratios $r(x)/x$ and $r'(x)/x$ (i.e., whose sum is unity) model how partitioning depends on measure. (The model, although cast in quantitative terms, is only a generic, qualitative model. Proper scaling would have to be introduced in specific cases to get practically useable results.)

To recapitulate: To get a convenient mathematical model, then, all we have to do is to shift the logarithm one unit to the left, which is accomplished by adding one unit to the argument, giving us $\log(1 + x)$, with the understanding that log means the natural logarithm, and that x is nonnegative.

Let is call the function $r(x)$ defined by $\log(1 + x)$, and let us call its complement $r'(x)$, that is, the function defined by $x - r(x)$. Thus:

$$r(x) \equiv \log(1 + x)$$

and

$$r'(x) \equiv x - r(x).$$

Now we are in a position to define the contrasting and competing ratios (or, percentages) and perceptions so often encountered. The relevant two fractions are $r(x)/x$ and $r'(x)/x$, and can often be interpreted as probabilities.

Let's consider an example: If x is the number of items in an urn (an urn being a container from which a random selection is made), then $r(x)/x$ is the probability that there is a significant difference between drawing with replacement and drawing without replacement. Notice that for low x, the probability is high, and for large x, the probability is low, precisely the kind of tug-of-war so often encountered in the world.

We will call the function defined by r(x) the 'partition' function.

Let's consider another example: the 'thing versus medium' dichotomy mentioned by Kurt Lewin in his book 'Principles of Topological Psychology' (pp. 115-117). He used the example of a hut that is being approached by someone. When that person is far away, the hut is a 'thing' that is being approached, but once the person reaches the hut and enters it, the hut becomes a 'medium' through which the person travels. This dichotomy is easily captured by the partition function. Let the entrance of the hut be positioned at the origin of the coordinate system, and let x be the distance of the person from the entrance to the hut. Then the degree to which the hut is a 'thing' is $r(1/x)/(1/x)$.

Let's consider yet another example: If x is the age of a discipline, then $r(x)/x$ is the fraction of its activity that is speculation, and $r'(x)/x$ is the fraction of its activity that is observation. (cf: "observation versus speculation" – Edna Heidbreder, Seven Psychologies, p. 14)

(For many more examples of the partition function, see Appendix 1.)

Chapter 1, Section 3: The psychological coordinate system

We are now in a position to propose a psychological coordinate system. Notice that the function defined by r'(x) 'hugs' the horizontal axis (the 'x-axis') in the small. Let us define h(x) to be synonymous with r'(x), the difference in name intended to be suggestive of its use as a horizontal coordinate. Let also introduce the inverse function of h(x), and call it v(x). Notice that v(x) 'hugs' the vertical axis in the small. So, in the small, this coordinate system of h(x) and v(x) looks similar to the ordinary Cartesian coordinate system of the first quadrant. However, in the large h(x) and v(x) converge to the line y = x, indicative of their essential equivalence – "a star by a star" – to use Swinburne's phrase.

We will take the horizontal psychological coordinate (that is, h(x)) to be technical (or, impersonal) progress, and the vertical psychological coordinate (that is, v(x)) to be social (or, personal) progress. To phrase it in negative terms, failure of h(x) to be large enough results in injury, and failure of v(x) to be large enough results in insult. It is usually the male that tends to h(x), and the female who attends to v(x). This explains why the male is 'simple' and the female is 'complex' – namely, because the male is identified by his coordinate h(x), that is, his physical coordinate and psychological coordinate are equal, whereas the female's psychological coordinate is v(x) while her physical coordinate is v(1/x). That is, a man does his work by leaving, but a woman does her work by arriving. Hence it is that a man's departure is deterministic, but his arrival is random, whereas a woman's arrival is deterministic, but her departure is random. For example, in the novel "Shane" by Jack Schaefer, the hero's arrival is random. (Indeed, the original title of the novel was "Stranger from Nowhere"). For another example, in the movie "Casablanca", Ilsa Lunt's arrival is deterministic, but her departure was random.

If, in regard to export (that is, production as opposed to consumption) the two distances from the origin to the two psychological coordinates are not equal, then the situation is not in equilibrium, and many stories consist of pointing out the (initial) lack of equilibrium and working towards (eventual) equilibrium by way of bringing the length of the shorter path up to equality of length of the other path. For example, the great charm and beauty of Cinderella was out of sync with her great poverty, but her poverty was eventually replaced by wealth matching her charm and beauty. Another example is Owen Wister's 'The Virginian'. The man's wealth, in the form of awesome competence, is eventually matched by acquiring a charming and beautiful bride. On the other hand, if the vertical coordinate exceeds the horizontal coordinate resume333

The partition function, with its associated psychological coordinate system, gives the general outline of what happens in regard to psychology, and the world in general. What happens in regard to the individual in regard to psychology is best given by abstract mathematical considerations, as Kurt Lewin correctly intuited.

Chapter 2: Application of abstract mathematics to psychology

Chapter 2, Section 1: Clearing the air

Let us now consider the description of the psychology of individuals via abstract mathematics. We being with topology. First, the term 'Topology' is an umbrella term, and is often used as an abbreviation for one of several sub-disciplines. The most basic sub-discipline, upon which all the others depend, is **General Topology**, and it is only this sub-discipline to which we refer herein by the term 'topology'. (An excellent introductory text for those interested is '**Introduction to Topology**' by **Bert Mendelson**.) General Topology is the Topology used by Kurt Lewin in his book '**Principles of Topological Psychology**'. Lewin displays a bit of amusing psychology in his use of Topology. He states (p. vii) that he at first tried to use **advanced topological concepts**, but found it too difficult to get worthwhile results, and so retreated to using only the most basic notions:

"After several attempts to employ the more complicated concepts of topology, I found it both sufficient and more fruitful to refer to the most simple topological concepts only."
-- Kur Lewin, Principles of Topological Psychology (p. vii)

What is amusing is that in his book he makes a glaring mathematical mistake. This raises the possibility that the real reason he couldn't make good use of advanced topics was that he simply didn't have the mathematical maturity to do so. Anyway, he then retreats to using only the very definition of a topological space. (Once burned, twice shy?) His mantra, that topology does not deal in distance, is also problematical. If it means that topology doesn't *need* the notion of distance, then that is certainly correct, but if it means (as Lewin seems to imply) that topology *necessarily* doesn't, and cannot, deal with distance, then it is certainly incorrect. The versions of topology dealing with distance are known as *metric spaces*. Indeed, historically, metric spaces were the *first* topological spaces to be studied. (The telltale mathematical mistake made by Lewin occurs on p. 137, in which he assumes that pairwise non-intersection is equivalent to an empty intersection of the entire collection. This mistake does not invalidate his approach, and is easily fixed, but it, and other quibbles, such as failing to use a comma in the expression B=f(PE), show his lack of fluency in mathematics.)

Retreating all the way back to the definition of a topological space, Lewin fails to make use of the most elementary tools offered by topology. For example, in order to define a path, he resorts to using a portion of a Jordan curve, even while admitting that it problematically (for the notion of a path) does not allow for self-intersection. But the definition of a path is very simple: a continuous mapping from the unit interval into the space in question. (Again, these criticisms of Lewin do not invalidate his approach. They merely point out the blemishes in his presentation.)

There are three levels of explanation: metaphor, analogy, and identity. In the literature, Lewin was criticized to using topology not to any novel effect, but merely as metaphor (similar to the criticism aimed at Furstenburg's 'topological' proof of the infinitude of primes). My take is that even if his entire use of topology was metaphorical, it at least served to point the way to a very fruitful line of development. In one of the sections of the book he mentions the situation of someone from afar approaching a hut (THING AND MEDIUM pp. 115-117). I think we can use this situation to discuss the contribution Lewin made in his book 'Principles of Topological Psychology'. Let us call the hut Topological Psychology, and the person approaching the hut Lewin. Now, I think we are all in agreement that Lewin at least reached the entrance to the hut. If he did not enter it, but merely peered inside, then the criticism that he used topology merely metaphorically would be valid. If he entered the hut and wandered around in it, that would correspond to a complete fulfillment of the claim implied by the title of the book. Then there is the possibility that he remained at the entrance, but did succeed in getting his foot in the door. I think a strong case could be made that he at least did this much. In any case, he himself admitted that his theory would need re-working and extension:

"The concepts that we here offer will certainly have to be revised in the course of time."
-- Kur Lewin, Principles of Topological Psychology (p. 7)

What we are going to do here is to use topology to give **analogies** for various situations and phenomena, without making any attempt to correlate or connect them with what Lewin does in his book, although in fact correlating and connecting with them when convenient or apropos. That is to say, because of his poor grasp of topology (or, lack of mathematical maturity), Lewin left a lot of low-hanging fruit to be plucked by others having (even slightly) greater mathematical maturity. Incidentally, aside from all mathematical considerations, Lewin's book (like Heidbreder's) is simply a pleasure to read due to the brilliance of the exposition of the concepts of psychology.

Let us note here some correlations of Lewin's concepts with those found elsewhere.

"Concepts which permit gradual transition between oppositions." p. 9
Besides use of the partition function, a nice analogy for this is that of the behavior of water above the critical point (in the p-v diagram), namely, that in that region water transitions continuously from liquid to vapor.

"In general, the descriptions that have been most valuable to science have not been made by scientific methods." (p. 13)
Perhaps the scientist is the 'cop', whereas the novelist is the 'robber'. (– the idea that progress is made by alternating between 'robber' and 'cop' – that is, the robber comes up with proposals, and the cop rejects those that are not viable)

"The center of interest shift from states to changes of state." (p. 16)
This is strongly similar – indeed isomorphic – to what happens in Differential Equations.
and a similar thought:
"one often realizes what the atmosphere has been only when it changes" (p.19)
and yet another similar thought:
"differential time sections" (p.25)
It is also worth pointing out that one can know changes of state without knowing states. Entropy is the canonical example of this: entropy states can be determined only up to an additive constant, but entropy differences are absolute, because that additive constant, whatever it is, disappears upon subtraction of two entropy levels.

"one must distinguish between "appearance" and "underlying reality"" (p.19)
This is similar to the "manifesting units" and "organizing units" mentioned by Heller and Macris in their book **'Parametric Linguistics'**.

"the philosophical Utopia of a universal science" p.21
What Lewin fails to appreciate is that psychology is the universal science: complete knowledge of the mind of man would amount to complete knowledge of the universe. (old joke:
A: Astronomy includes all the stars in the sky! B: And psychology includes all astronomers.)
Even Lewin, forgetting himself, seems to agree:
"the main purpose of psychology is ... to explain reality" (p.22)
Notice that he did not say "psychological reality", but simply "reality" – so, universal science!

"Such an understanding, if any, needs the cooperation of a group." (p. viii)
Nothing supports 'cooperation of a group' like Esperanto. So, let's all learn Esperanto and get on with the work!

"The confusion of "objective" with "physical" and of "logically general" with "equal for all" has led to grave conceptual and methodological errors in psychology." (p. 25)
This is a brilliant insight on the part of Lewin. Perhaps the problem he is pointing out here should be upgraded to having its own name among the Wikipedia list of fallacies.

"The confusion of historical and systematic concepts and problems is an essential characteristic of pre-Galilean or Aristotelian thinking of a period in psychology which is now coming to an end and which has led to momentous errors." (pp. 31-32)
Again, this is a brilliant insight on the part of Lewin. However, I wonder to what extent Lewin is using Galileo and Aristotle as merely convenient milestones. After all, Galileo made major mistakes, too. Lewin is attempting to correct a meta-mistake here, and that seems to be a very difficult thing to do, as there seems to be a strong universal tendency to turn a blind eye to meta mistakes, that is, an expert in a given domain seems to suffer no loss of prestige by making a whopper of a meta mistake. (So, we could call this phenomenon 'meta-mistake mongering'.) For example, David Hilbert made meta mistakes in his famous list of 23 problems, but this does not seem to have hurt his stature as a mathematician in the least. Galileo also made a meta mistake in regard to dropping the two different weights from the tower of Pisa. He concluded that they fell at the same rate. In fact, according to Newtonian physics, they do not. The conclusion that he should have drawn was "If there was a difference in their times of fall, it was too small for us to detect with our present instruments." (Galileo also gave the wrong answer to the brachistochrone problem.)

"Unclear zones of unsharp transitions lead more often to tension and conflicts." (p. 122)
Indeed. As Robert Frost said, good fences make good neighbors.

"whether it is the life situation or the momentary situation which comes more strongly into the foreground" (p.24)
General Patton, in his famous saying, was addressing when the momentary situation is more strongly in the foreground ("A good plan executed now is better than the perfect plan executed next week.")

"only the present situation can influence present events" (p.34)
Lewin does not mention Markov anywhere in his book, but his presentation would have been stronger had he done so, because he was re-inventing the Markovian wheel, so to speak. (A Markov process is one in which the future state is determined solely by the present state – not by historical states.)
and a similar thought:
"Past events can only have a position in the historical causal chains whose interweavings create the present situation." (p.35)

"periods of apparently continuous transformations are followed by periods of crisis with sudden changes of structure" (p.36)
cf: Stephen Jay Gould's thesis of discontinuous evolution

"the dynamic structure of a situation is not an immediately given fact" (p. 82)
Don't companies exploit this fact in order to pass on hidden costs to the consumer? For example, it is a deep defect for an apartment complex not to have an on-site manager, because it takes an on-site manager to correctly handle the 'momentary situations' (to use Lewin's terminology) that inevitably arise, but the lack of an on-site manager is not obvious to the casual observer, including prospective tenants.

"A complete representation of one situation would mean that the whole task of psychology is completed." (p. 82)
This is analogous to what happens with analytic functions in the theory of Complex Variables: If you have complete information about a given analytic function on any nonempty open set, you have complete information about the given analytic function throughout its domain.
If the whole task of psychology is completed, then so is the whole task of physics, and every other individual science, because, as we pointed out at the outset, a complete theory of psychology would amount to a Theory of Everything, given its root position among the sciences.
also:
"the investigation and representation of each single case is an infinite task in itself" (p.17)
(one of the messages of religion? – e.g., 'nitpicking and hair-splitting' – regarded as necessary – regarding the Torah)

Chapter 2, Section 2: The indiscrete topology

Our first analogy is that of modeling any phenomenon, such as a dust explosion, in which a total change follows from an infinitesimally small disturbance. A wrong glance in a certain sensitive situation would be a psychological example, or, to use an old joke, "New York now numbers the most number of people that you shouldn't make a sudden move around." The instrument of the analogy is simply that of the indiscrete topology of a given set of large cardinality. The explosion consists of what happens when you take the closure of any singleton, with the interpretation that taking the closure of a set constitutes 'consuming' the set.

The analogy of the indiscrete topology also closely mimics the well-known phenomenon of the tail wagging the dog, and jokes based on this principle. For example: Q: How does a spoiled rich girl change a lightbulb? A: She says, "Daddy, I need a new apartment."

Another phenomenon modeled by the indiscrete topology is that of the situation of 'in for a penny, in for a pound'.

Another phenomenon modeled by the indiscrete topology is the pulling of the trigger of a firearm.

Another phenomenon modeled by the indiscrete topology is the application of the 'non-compete' clause of the agreement that even temp / part-time workers have to sign. (very similar to 'in for a penny, in for a pound')

At the other end of the spectrum of possible topologies is the discrete topology. Movement towards it constitutes progress, the movement away from routine killing of the relatives of one's enemies. Winwood Reade, in his book 'The Martyrdom of Man', describes this gradual abandonment of blood feud on p. 88:

"In a higher state of society this family system disappears; individualism becomes established."

Nothing shouts individualism more than the discrete topology.

This inevitable / required movement towards individualism also provides a plot device for the novelist to easily impute evil to a certain character, as was done in the movie "Ocean's Eleven" (2001 version), when with a brief bit of dialog it is revealed to the audience that the casino owner takes revenge on his enemies relatives. Clearly, the casino owner never got the memo.

Chapter 2, Section 3: Palpable boundaries

Practical matters usually involve some kind of measure, even when topological considerations are at the forefront. Of course, it may happen that the two are combined, as in the case of a metric space, but it can, and often does, happen that the measure regime and the topological regime coexist in parallel. Introducing measure might seem like a heavy assumption, but Lewin makes an implicit assumption no less heavy when he willy-nilly drops Jordan curves into the discussion involving "purely topological" concerns. The introduction of Jordan curves forces all subsequent discussion to the plane, which is rich in measure, to say nothing of metric.

So, we will now introduce topology, along with measure and some numerical notions superimposed on the given topological regime. Once this is done, we can introduce the concept of 'palpable' boundaries, and generalize the situation of the indiscrete topology to 'thick boundaries', as a special case of palpable boundaries, which will allow us to describe a much wider set of phenomena. The indiscrete topology is obviously a special case of that – so special that no extra numerical notions are needed to describe it.

Abstract Formulation of Psychological Dynamics

"There is nothing so practical as a good theory."
-- Kurt Lewin

This development is to be understood as taking place within **Zermelo-Fraenkel set theory**.

Axiom 1. Ξ is an infinite set.

Definition 1. A member of Ξ is called a **monad**.

Remark 1. This may or may not be a monad in the sense of **Leibniz**.

Axiom 2. **Quixote** is a set of ordinals.

Axiom 3. Every **nonnegative integer** is an element of Quixote.

Axiom 4. $\omega \in$ Quixote.

Remark 2. ω is, as it is usually defined, the first transfinite ordinal.

Definition 2. A **quordinal** is an element of Quixote.

Remark 3. The user of this system specifies which ordinals are quordinals, with the only requirement being that every nonnegative integer be a quordinal and that ω be a quordinal.

Definition 3. For each quordinal n, $X_n = \{(n,x) \mid x \in \Xi\}$.

Axiom 5. For each quordinal n, (X_n, T_n) is a topological space.

Remark 4. (X_n, T_n) describes things as they **are**.

Axiom 6. For each quordinal n, (X_n, W_n) is a topological space.

Remark 5. (X_n, W_n) describes things as they **ought to be**. (cf: 'The Pure Theory of Law' by Hans Kelsen)

Remark 6. cf: the concept of a **bi-topological** space.

Axiom 7. For each quordinal n, (X_n, Π_n, μ_n) is a measure space.

Remark 7. Not all subsets of X_n are measurable, but in this work we will, for the sake of expository convenience, assume that all subsets of X_n encountered, in definitions and elsewhere, are in fact measurable.

Definition 4. **The_Da_Vinci_Staircase** = $\{X_n \mid n \text{ is a quordinal}\}$.

"A work of art is never completed, merely abandoned."
-- Leonardo da Vinci

Definition 5. **the_world** = $\cup$The_Da_Vinci_Staircase.

Axiom 8. Λ is a set having no connection with Ξ.

Definition 6. An **ur-element** is an element of Λ.

Remark 8. The ur-elements are used for **referencing and indexing**. Note that having no connection with Ξ is a much stronger statement than merely that it does not intersect Ξ. It means also that it does not intersect the power set of Ξ, nor the power set of the power set of Ξ, and so on, and that it does not intersect X_n for any n, and so on.

Axiom 9. **Entities** is a subset of Λ.

Definition 7. A member of **Entities** is called an 'entity of interest' (or, simply 'entity' for short).

Remark 9. An entity is anything that can be conceived as having a unified and independent existence. An entity of interest is any entity we wish to consider. An entity of interest could be a person, a group of people, a group of groups of people, and so on, and the same goes for an animal, a plant, and a chemical compound or element, and so on. Not all entities are of interest to us in this development, however, for example, monads.

Axiom 10. Orion is a (possibly empty) set of topological connectedness axioms.

Remark 10. The user of this system chooses what constitutes the membership of **Orion**. Some examples are: connectedness, path-connectedness, arcwise-connectedness, hyperconnectedness, and ultraconnectedness. It is permitted for Orion to be empty.

Definition 8. For each quordinal n, an **Orion subset of X_n** is a subset M of X_n such that M satisfies each and every member of Orion.

Definition 9. "$\forall(m,n,t)$" is an abbreviation for "For each entity m, for each quordinal n, and for each point of time t".

Axiom 11. $\forall(m,n,t)$, each of $v_1(m,n,t)$ and $v_2(m,n,t)$ is an Orion open subset of X_n having strictly positive, but finite, measure (with respect to μ_n, of course) and a nonempty (topological) boundary. $v(m,n,t)$ is called the **nucleus** (synonymously, the **core**) of [m,n,t].

Convention 1. A subscript of 1 refers to how something **is**, and a subscript of 2 refers to how something **ought to be**.

Axiom 12. $\forall(m,n,t)$, each of $\zeta_1(m,n,t)$ and $\zeta_2(m,n,t)$ is a (topologically) closed subset of the (topological) exterior of $v(m,n,t)$ and superset of the (topological) boundary of $v(m,n,t)$. $\zeta(m,n,t)$ is called the **periphery** (synonymously, the **cytoplasm**) of [m,n,t].

Convention 2. Absence of a subscript means that the statement is true for both subscripts.

Theorem 1. $\forall(m,n,t)$, $v(m,n,t)$ is nonempty.
Proof: The measure of the empty set is 0. ∎

Theorem 2. $\forall(m,n,t)$, $\zeta(m,n,t)$ is nonempty.
Proof: By Theorem 1, the boundary of $v(m,n,t)$ is nonempty, and by Axiom 12, the boundary of $v(m,n,t)$ is a subset of $\zeta(m,n,t)$. Hence $\zeta(m,n,t)$ is nonempty. ∎

Axiom 13. $\forall(m,n,t)$ and for each x in $\zeta(m,n,t)$, x is reachable from $v(m,n,t)$.

Remark 11. In other words, the periphery is a subset of the set of points reachable from the core.

Remark 12. The core supplies the periphery with connectedness, and the periphery supplies the core with separateness (by acting as a buffer vis-à-vis other entities).

Definition 10. $\forall(m,n,t)$, n is called the **irreality level** of $v(m,n,t)$.

Definition 11. "[m,n,t]" is an abbreviation for "the entity m at irreality level n at time t".

Remark 13. Reality is, of course, irreality of level 0. The term 'irreality' is taken from **Lewin**.

Remark 14. Irreality level 0 is the **wall of shadows** of Plato's cave, and irreality level ω constitutes the objects whose shadows are thrown upon the wall. The whole point of the Roman Catholic doctrine of **transubstantiation** is to suggest the existence of irreality level ω to those whose habitual / exclusive focus is on irreality level 0. There is a brilliant passage in C. S. Lewis's 'The Screwtape Letters' that captures this idea very well:

"an unalterable conviction that, whatever odd ideas might come into a man's head when he was shut up alone with his books, a healthy dose of "real life" (by which he meant the bus and the newsboy) was enough to show him that all "that sort of thing" just couldn't be true."

When Pascal talks about sitting quietly in a room, he is referencing the prioritization of activity in irreality level ω:

"All the trouble in the world is due to the inability of a man to sit quietly in a room."
-- Blaise Pascal

This sentiment is echoed in Max Ehrmann's 'Desiderata': "Avoid loud and aggressive persons, they are vexations to the spirit."

Women start at irreality level ω, and descend, possibly too far. Men start at irreality level 0, and ascend, possibly not far enough. This is captured by the proverb, "A man is as good as he has to be. A woman is as bad as she dares to be." It also explains the felt truth behind Lady Astor's quote: "I married beneath me. All women do."

Convention 3. For the sake of avoiding cumbersome phraseology, '**irreality**', unless otherwise specified or implied, is a reference to an irreality level > 0, and '**reality**' means irreality level 0.

Remark 15. The popular term '**sword and sorcery**' refers to the (personal) struggles within reality (symbolized by 'sword') and the (personal) struggles within irreality (symbolized by 'sorcery'). As the Wikipedia article on sword and sorcery states, tales of sword and sorcery, "though dramatic, focus mainly on personal battles rather than world-engaging matters". So, 'sword and sorcery' herein could be taken as a synonym for The_Da_Vinci_Staircase, or as a synonym for what happens therein. From the viewpoint of the state, all this is referred to as '**bread and circuses**', with 'bread' standing for satisfaction of concerns in reality, and 'circuses' standing for satisfaction of concerns in irreality.

Axiom 14. $\forall$(m,n,t), **urgency(m,n,t)** is a nonnegative real number, and urgency(m,n,t) $\geq$ urgency(m,n+1,t).

Axiom 15. For each entity m, and for each point of time t, there exists a quordinal n such that urgency(m,n,t) > 0.

Remark 16. No one is exempt from care.

"MISFORTUNE, n. The kind of fortune that never misses."
-- Ambrose Bierce, The Devil's Dictionary

Axiom 16. For each entity m, and for each point of time t, for each quordinal p and for each quordinal q, if $p < q$ and urgency(m,p,t) > 0, then urgency(m,q,t) > 0.

Remark 17. Urgency at a lower level implies urgency at all higher levels. (cf: "A poet can survive anything, except a misprint.")

Axiom 17. For each entity m, and for each point of time t, **activity_level(m,t)** is a quordinal, called the **activity level** of entity m at time t.

Definition 12. For each entity m, and for each point of time t, m is said to be **ridiculous** at time t fif there exists a quordinal n such that urgency(m,n,t) > 0, and activity_level(m,t) $> n$.

Remark 18. A **ridiculous** entity is one whose activity level is higher than a level having urgency measure > 0. (The political left easily falls into this error.)

Definition 13. For each entity m, and for each point of time t, m is said to be **contemptible** at time t fif urgency(m,activity_level(m,t),t) $= 0$.

Definition 14. A **contemptible** entity is one whose activity level has an urgency measure of 0. (The political right easily falls into this error.)

Remark 19. The idea of the Sabbath (and of the forgiveness of debts every 7 years) is that of giving some breathing-room for making a serious effort to raise the active level. For the priesthood, the active level > 1, typically, or presumably, $\gg 1$.

Remark 20. "for each point of time", a phrase that will often appear in definitions and axioms, does not mean all possible points in time, but only those that pertain to the situation in question. So, for example, if something comes into existence at time 5 and ceases existence at time 14, then the points in time are the elements of [5,14).

Axiom 18. For each quordinal n, **Linear_Linkage$_n$** is an equivalence relation on X_n.

Definition 15. $\forall$(m,n,t) and for each x in X_n, the statement that x is **reachable** from v(m,n,t) means that Linear_Linkage(x,y) for some point y in v(m,n,t).

Axiom 19. $\forall$(m,n,t), and for each x in ζ(m,t), $\boldsymbol{\eta(v(m,n,t),x)}$ is a nonnegative real number.

Remark 21. So, by the absence-of-a-subscript convention, we have both η_1 and η_2.

Remark 22. Intuitively, $\eta(v(m,n,t),x)$ is how far x is from $v(m,n,t)$. By means of this metric we can distinguish between the '**near**' points of the periphery and the '**far**' points of the periphery, which have differing psychological features.

Axiom 20. $\forall(m,n,t)$, for each x in $v(m,n,t)$, $\eta(v(m,n,t),x) < 0$.

Remark 23. Similar to how the natural logarithm behaves on $(0,1)$.

Definition 16. $\forall(m,n,t)$, $v(m,n,t)$ is said to have a **palpable periphery** fif there exists a point x of $\zeta(m,n,t)$ such that $\eta(v(m,n,t),x) > 0$.

Axiom 21. $\forall(m,n,t)$, and for each x and y in $\zeta(m,n,t)$, there exists a path f from x to y such that g is a **continuous function**, where g is a function whose domain is $[0,1]$ such that for each h in $[0,1]$, $g(h) = \eta(v(m,n,t),f(h))$.

Remark 24. That is, distance from the core varies smoothly.

Definition 17. $\forall(m,n,t)$, $v(m,n,t)$ is said to have a **thick periphery** fif there exists a point x of $\zeta(m,n,t)$ such that $\eta(v(m,n,t),x) > 1$.

Remark 25. An archetypal case of thick periphery is that of unnecessarily large footprints, that is, when something consumes more than is needed. A canonical example is that of the remark of Blaise Pascal in a letter: "I'm sorry this letter is so long. I didn't have the time to make it shorter."

Remark 26. A frequently encountered large footprint is when putting leftovers into the refrigerator. If you don't put the leftovers into a smaller dish, you are unnecessarily taking up valuable space in the refrigerator.

Remark 27. Another example of a large footprint is what users experience when interacting with a computer system that was not designed to handle all contingencies. For example, if I order a rubber stamp from an online vendor, I later get an email containing the line: "Please allow 5-7 business days for delivery unless you selected a premium delivery option." But they know perfectly well whether I've selected a premium delivery option. Instead of handling that contingency separately, they send this one-size-fits-all email message. Indeed, 'one-size-fits-all' is the mantra of a certain class of large footprints.

Remark 28. The most notorious example of a large footprint is, of course, that of 'collateral damage'. Another example of a large footprint, which is the dual / poetic justice version of 'collateral damage', consists of logging in to the system with higher authority than you need, and then accidentally doing great damage to your system. (Hence the proverbial admonition of employing minimal sufficiency.)

Remark 29. Another example of a large footprint is the phenomenon referred to by the saying: "If you give him an inch, he'll take a mile."

Definition 18. $\forall(m,n,t)$, $v(m,n,t)$ said to have a **thin periphery** fif for each x in $\zeta(m,n,t)$, $\eta(v(m,n,t),x) < 1$.

Axiom 22. $\forall(m,n,t)$, if $v(m,n,t)$ has a thick periphery, then $v(m,n,t)$ is **arcwise-connected**. (cf: Virtue is its own reward.)

Axiom 23. $\forall(m,n,t)$, η attains a maximum on $\zeta(m,n,t)$. ($\zeta(m,n,t)$ is said to be '**of finite thickness**'.)

Remark 30. Roughly speaking, no matter how capable you are, your **bandwidth** is finite.

Remark 31. So, informally, '**bandwidth**' could be taken as synonymous with the measure of the periphery, or, in the most extreme case, as the measure of the union of the core and the periphery.

Definition 19. $\forall(m,n,t)$, the **membrane** of $v(m,n,t)$ is the set elements x of $\zeta(m,n,t)$ such that $\eta(v(m,n,t),x) = 0$.

Definition 20. $\forall(m,n,t)$, the **property line** of $v(m,n,t)$ at time t is the set of elements x of $\zeta(m,n,t)$ such that $\eta(v(m,n,t),x)$ is maximum.

Theorem 3. If the property line is equal to the membrane, then the periphery is thin.
Proof: Obvious. ∎

Convention 4. '**strongly separated**' is a user-defined term, relative to a specific application of this system.

Remark 32. For example, the user might specify that it means that the set in question, when considered as a subspace, is Hausdorff. Or again, that the set in question has a subset H of measure > 1 such that H, when considered as a subspace, is normal. And so on.

Definition 21. A **resource** is an entity whose core is strongly separated and whose periphery is thin.

Definition 22. **Buried treasure** is an entity whose core is strongly separated but whose periphery is thick. (i.e., a "hidden" resource)

Definition 23. A **dependent** is an entity whose core is connected but whose periphery is thin.

Remark 33. cf: "naked among wolves"

Definition 24. An **independent entity** is one whose core is connected and whose periphery is thick.

	periphery thin	periphery thick
core connected	a dependent	an independent entity
core strongly separated	a resource	buried treasure

Definition 25. $\forall$(m,n,t), $\boldsymbol{\pi(m,n,t)}$ = v(m,n,t) $\cup\zeta$(m,n,t), and is called the **personal space of m at irreality level n at time t**.

Definition 26. For each entity m, and for each point in time t, $\boldsymbol{\Omega(m,t)}$ = {π(m,n,t) | n is a nonnegative integer}.

Definition 27. For each entity m, and for each point in time t, **personal_space(m,t)** = $\cup\Omega$(m,t).

Remark 34. 'personal space' is not to be confused with 'life space' – which will be defined presently.

Axiom 24. $\forall$(m,n,t), $\boldsymbol{e(m,n,t)}$ is a subset of Λ, and is called the **environment** of π(m,n,t).

Axiom 25. For each entity m, and for each point in time t, $\boldsymbol{E(m,t)}$ = {e(m,n,t) | n is a nonnegative integer}., and $\cup$E(m,t) is called the **environment of entity m at time t**.

Definition 28. For each entity m, and for each point of time t, **environment(m,t)** = $\cup$E(m,t), and is called the **environment of entity m at time t**.

Definition 29. For each entity m and for each point in time t,
life_space(m,t) = (life_space(m,t),environment(m,t)), and is called the **life space of m at time t**.

Remark 35. Notice that, unlike Lewin, we don't try to mix oil and water, that is, we don't regard the person (personal space) and the environment of the person as being all-one-thing.

Remark 36. Documentation (especially map-making) corrals momentary situations into the life situation. Corraling certain momentary situations into the life situation might constitute a wildly important goal.

Convention 5. The measure of a subset H of the personal space of an entity m at time t will be denoted generically by $\boldsymbol{\mu(H)}$, with it understood that the user will supply the subscript for μ as appropriate.

Axiom 26. saint_threshold is a real number ≥ 1.

Definition 30. For each entity m, and for each point of time t, m is said to be a **saint** at time t fif $\mu(\zeta$(m,n,t))/μ(v(m,n,t)) > saint_threshold.

Definition 31. For each entity m, and for each point of time t, m is said to be a **desperado** at time t fif $\mu(\zeta$(m,n,t))/μ(v(m,n,t)) < 1.

Axiom 27. **heavy_hitter_threshold** is a real number > 1.

Definition 32. For each entity m, and for each point of time t, m is said to be a **heavy hitter** at time t fif μ(life_space(m,t)) > heavy_hitter_threshold.

Remark 37. A **saint** is someone whose core is small relative to his periphery. A **desperado** is someone whose core is large relative to his periphery. A **heavy hitter** is someone whose life space has large measure. Many **stories** are about those who are gradually revealed to be not only saints, but heavy hitters (e.g., Shane, The Magnificent Seven, The Virginian).

Remark 38. You can expand your core to some degree (by puffing yourself up), and thereby increasing the measure of your life space, perhaps even to the point of becoming a heavy hitter, but at the cost of losing some of the connectedness within your core, connectedness that remains lost even if you later reduce your core to its former size. Repeated episodes of this lead to cognitive decline.

Axiom 28. **Emissaries** is a subset of Λ.

Definition 33. An **emissary** is a member of Emissaries.

Definition 34. For each entity m and for each point of time t, **Incoming(m,t)** is a (possibly empty) subset of Λ.

Definition 35. For each entity m, for each point of time t and for each element z of Incoming(m,t), **source(m,t,z)** is an entity.

Definition 36. For each entity m_1, for each entity m_2, for each point of time t, and for each emissary z, **launch(m_1,m_2,t,z)** fif $z \in$ Incoming(m,t), and m_2 = source(m_1,t,z).

Remark 39. m_1 might equal m_2, as when someone mails themselves a letter.

Axiom 29. For each emissary z, **γ(z)** is a real number.

Definition 37. For each emissary z, z is said to be **exothermic** fif γ(z) > 0.

Definition 38. For each emissary z, z is said to be **endothermic** fif γ(z) < 0.

Definition 39. For each emissary z, for each entity m, and for each point of time t, z is said to be a **gift** for m at time t fif $z \in$ Incoming(m,t), and z is exothermic.

Axiom 30. For each emissary z and for each point of time t, **ψ(z,t)** is an element of the_world.

Remark 40. ψ(z,t) is the **impact point** of z, at time t.

Definition 40. For each entity m and for each point of time t, an **impingement on m at time t** is an endothermic emissary z such that $\psi(z,t) \in \cup\Omega(m,t)$.

Definition 41. For each entity m and for each point of time t, **whirlwind(m,t)** = {z | z is an impingement on m at time t}.

Remark 41. cf: the terminology of **FranklinCovey**.

Axiom 31. For each entity m and for each point of time t, **goals(m,t)** is a (possibly empty) subset of Λ.

Axiom 32. For each entity m and for each point of time t, **wigs(m,t)** is a (possibly empty) subset of goals(m,t).

Remark 42. 'wigs' stands for '**wildly important goals**'.

Remark 43. cf: the terminology of **FranklinCovey**.

"If you want to build a ship, don't drum up the men to gather wood, divide the work, and give orders. Instead, teach them to yearn for the vast and endless sea."
-- Antoine de Saint-Exupéry

Definition 42. For each entity m and for each point of time t, **core(Ω(m,t))** = $\cup${M | M is the core of some member of Ω(m,t)}.

Definition 43. For each entity m and for each point of time t, **periphery(Ω(m,t0)** = $\cup${M | M is the periphery of some member of Ω(m,t)}.

Definition 44. For each entity m and for each point of time t, an **injury** to entity m at time t is an impingement on m at time t that is a member of core(Ω(m,t)).

Definition 45. An injury is said to be **minor** fif $\gamma > -1$. (i.e., endothermic, but not much so, that is neither as wide as a church door, nor as deep as a well)

Definition 46. An injury is said to be **major** fif $\gamma < -1$. (i.e., very endothermic)

Axiom 33. For each separation axiom A of General Topology, **w(A)** is a positive real number (the 'weight' of the axiom).

Axiom 34. If each of A and B is a separation axiom of General Topology, and A implies B, then $w(A) \geq w(B)$.

Remark 44. So, the more a separation axiom implies, the greater weight it is allowed to have, but we allow for the possibility that for some application of this system all axioms be given the same weight.

Axiom 35. **J** is a set of separation axioms of General Topology such that no member of J implies any other member of J.

Remark 45. The user of this system determines what constitutes the members of J, and what their weights are.

Remark 46. The purpose of the non-implication requirement is to give rise to a well-defined partition.

Remark 47. The non-implication requirement can have some surprising results. For example, as Ralph Kopperman mentions in his article 'Asymmetry and duality in topology', a normal space is not necessarily a T_1 space.

Definition 47. Suppose that n is a nonnegative integer and M is an infinite subset of X_n. Let K be the subset of axioms A of J for which there exists an infinite subset H of M such that H, considered as a subspace, satisfies A. Then if K is empty, then the **degree of tractability of M** is 0. If K is not empty, then let P be the partition of M induced by K (in the obvious way). Then the **degree of tractability of M** is the sum of the products $\mu(H)w(A)$, where the sum is taken over A in K and H in P (in the obvious way).

Theorem 4. If n is a nonnegative integer and each of M and H is a subset of X_n and M is a subset of H, then the degree of tractability of H is no less than the degree of tractability of M.
Proof: Obvious. ∎

Remark 48. In other words, the larger a set is, the easier it is for it to be tractable. (This is why, in the novel 'Shane', the rancher wanted to buy out the settlers. Shane himself remarked this to Joe Starrett.)

Axiom 36. $\forall(m,n,t)$, **$\$(v(m,n,t))$** is a nonnegative real number, called the **earning potential** (synonymously, the **economic potential**) of $v(m,n,t)$.

Axiom 37. $\forall(m,n,t)$, $\$(v(m,n,t)$ is an increasing function of the degree of tractability of $\zeta(m,n,t)$. That is, if f(t) = the degree of tractability of $\zeta(m,n,t)$ and g(t) = $\$(v(m,n,t))$, then if f is increasing, then so is g.

Definition 48. $\forall(m,n,t)$, the **strength** of $v(m,n,t)$ is equal to the reciprocal of the amount of time required for it to acquire a periphery double the size of its current periphery.

Definition 49. $\forall(m,n,t)$, the statement that $v(m,n,t)$ is **magnificent** means that its strength is > 1. (cf: Shane, The Magnificent Seven, The Virginian)

Axiom 38. **Topics** is a subset of Λ.

Definition 50. A **topic** is a member of Topics.

Axiom 39. **Deductive Mathematics** is a topic.

Axiom 40. **Monotheistic Theology** is a topic.

Axiom 41. **Concepts_and_Procedures** is a subset of Topics.

Definition 51. A **concept_or_procedure** is a member of Concepts_and_Procedures.

Remark 49. Flip-flop between 'and' and 'or' is itself an interesting psychological phenomenon. Externally, things referenced by 'and' are referenced internally by 'or'. To take another example, if the boss says to the computer programmer, "Give me a list of all our salesreps that live in Florida and Texas, the programmer passes the salesrep file using the logic 'if residence = Florida or residence = Texas, then print this salesrep'. This is a special case of what is referenced externally as being 'solid' and what is referenced internally as being 'liquid'. Another example of this phenomenon is the switch of positions of the national and provincial flags in the auditorium of a school vis-à-vis in the classrooms (the national flag being to the speaker's right in the auditorium, but to the students' right in the classroom).

Definition 52. For each entity m and for each point of time t, $\kappa(m,t) = \{f \mid$ there exists a topic h and there exists a point x in $\nu(h,n,t)$ and there exists a y in $core(\cup\Omega(m,t))$ such that Linear_Linkage(x,y)}. $\kappa(m,n,t)$ is the **knowledge** of entity m at time t.

Definition 53. For each entity m and for each point of time t, $\eta(m,t) = \{f \mid f \notin \kappa(m,t)$ and there exists a point of time $s < t$ such that $f \in k(m,s)\}$. $\eta(m,t)$ is the **dust bin** of m at time t (synonymously, **what m has forgotten** by time t).

Remark 50. We can now model the following situation:
A: "Do you know about etymology?"
B: "I have forgotten more about etymology than you know about etymology."
What B is asserting is: $\mu(\eta(B,t)) > \mu(\kappa(A,t))$.

Axiom 42. For each z in Λ and for each point of time t, $\tau(z,t) \in [0,1]$.

Definition 54. For each z in Λ, z is said to **be in foreground** at time t fif $\tau(z,t) = 1$.

Definition 55. If K is a nonempty subset of Λ and t is a point in time, then K is said to **be in foreground** fif for each x in K, x is in foreground.

Definition 56. If m is an entity and a and b are two points in time such that $a < b$, and $\varphi(m,t)$ is a constant z for all t in [a,b], then the **amount of concentration of m over the interval [a,b]** is equal to $(b - a)/h$, where h is the common value of $\sigma(z,t)$ for t in [a,b]. (For example, h could be taken to be $\sigma(z,a)$.)

Axiom 43. For each fixed point of time t, $d(m_1,m_2,t)$ is a metric on Entities.

Definition 57. For each point of time t, and for each two distinct entities m_1 and m_2, m_1 and m_2 are said to be **docked** (with each other) (synonymously, **in proximity** (to each other)) (synonymously, **close together**) at time t fif $d(m_1,m_2,t) < 1$.

Remark 51. Docking may be intentional, or may be accidental, as on a public conveyance, or may be arranged, such as assigned seating on an airplane.

Definition 58. For each point of time t_1 and for each point of time t_2 such that $t_1 < t_2$, for each two distinct entities m_1 and m_2, $[t_1,t_2]$ is said to be a **docking episode** for m_1 and m_2 fif $[t_1,t_2]$ is a closed interval of maximal size such that for each t in $[t_1,t_2)$, m_1 and m_2 are docked (with each other) at time t, and m_1 and m_2 are not docked (with each other) at time t_2.

Remark 52. Velikovsky, in his book '**Worlds in Collision**', claims that the Earth has had docking episodes with other planets within recorded history.

Lewin does not use the term 'docking', but uses the concept without naming it.

As Lewin notes, control over docking is a major portion of one's psychological activity:

"Political struggles as well as struggles between individuals are nearly always struggles over the boundary of the space of free movement."
-- Kurt Lewin, Principles of Topological Psychology (p. 47)

Remark 53. Of course, the road to such struggles is paved, so to speak, with instances of large footprints. That is, an alien employing a large footprint inadvertently – or perhaps deliberately – 'steps on' the subject, that is, uses part of the life space of the subject (advancing economically depending on posting to foreground, and posting to foreground depending on footprints).

Remark 54. The thwarting of unwanted docking is a major concern. Most occurrences of docking, wanted or unwanted, are via language. This means that unwanted docking can often be nipped in the bud by not confirming commonality of language. If you're bilingual, you can respond to an unwanted request from a stranger (such as an aggressive panhandler) by switching to that other language. By the way, there's no faking it. Trying to make linguistically plausible-sounding utterances newly-minted by you on the spot under stress is next to impossible.

Remark 55. Pretending not to speak the language of the would-be docker is a special case of the more general idea of feigning lack of comprehension. For example: "My mother is far too clever to understand anything she doesn't like. -- Arnold Bennett

Definition 59. For each entity m_1, **Farm(m_1)** = {$(m_2,[t_1,t_2])$ | each of t_1 and t_2 is a point in time such that $t_1 < t_2$ and m_2 is an entity distinct from m_1 such that $[t_1,t_2]$ is a docking episode for m_1 and m_2}.

Theorem 5. For each two distinct entities m_1 and m_2, $(m_2,[t_1,t_2]) \in$ Farm(m_1) fif $(m_1,[t_1,t_2]) \in$ Farm(m_2).
Proof: Obvious. ∎

Axiom 44. For each entity m_1, and for each member $(m_2,[t_1,t_2])$ of Farm(m_1), **ι(m_1,(m_2,[t_1,t_2]))** is an element of [0,1].

Remark 56. $ι(m_1,(m_2,[t_1,t_2]))$ is the extent to which entity m_1 **initiated** the docking episode $[t_1,t_2]$ for m_1 and m_2.

Axiom 45. If m_1 and m_2 are two distinct entities, then **ι(m_1,(m_2,[t_1,t_2])) + ι(m_2,(m_1,[t_1,t_2])) ≤ 1.**

Remark 57. In case $ι(m_1,(m_2,[t_1,t_2])) + ι(m_2,(m_1,[t_1,t_2])) < 1$, a **third party** also contributed to initiating the docking episode, for example, assigned seating on an airplane.

Axiom 46. For each entity m_1 and for each member $(m_2,[t_1,t_2])$ of Farm(m_1), if $ι(m_1,(m_2,[t_1,t_2])) > 0$, then **ρ($m_1$,($m_2$,[$t_1$,$t_2$]))** is an element of the (topological closure of) the first quadrant of the complex plane.

Remark 58. $ρ(m_1,(m_2,[t_1,t_2]))$ is the **amount and kind of right (or reason)** by which entity m_1 initiated, to whatever extent he did, the docking episode $[t_1,t_2]$ for m_1 and m_2. The horizontal component is the right (or reason) based on his **momentary situation**, and the vertical component is the right (or reason) based on his **life situation**.

Definition 60. If $|ρ(m_1,(m_2,[t_1,t_2]))| < 1$, then it is said that m_1 acted with **insufficient right (or reason)** in initiating, to whatever extent he did, the docking episode $[t_1,t_2]$ with entity m_2.

Definition 61. If $|ρ(m_1,(m_2,[t_1,t_2]))| > 1$, then it is said that m_1 acted with **sufficient right (or reason)**, to whatever extent he did, the docking episode $[t_1,t_2]$ with entity m_2.

Definition 62. For each entity m_1 and for each member $(m_2,[t_1,t_2])$ of Farm(m_1), m_1 is said to have committed an **intrusion** regarding the docking episode $[t_1,t_2]$ of m_1 and m_2 fif $ι(m_1,(m_2,[t_1,t_2])) = 1$ and $ρ(m_1,(m_2,[t_1,t_2])) < 1$.

Axiom 47. For each entity m_1 and for each member $(m_2,[t_1,t_2])$ of Farm(m_1), **Involvement(m_1,(m_2,[t_1,t_2]))** is a nonempty set, each element of which is a nonnegative integer.

Remark 59. Involvement(m_1,(m_2,[t_1,t_2])) is the set of **irreality levels** of m_1 involved in the docking episode $[t_1,t_2]$ of m_1 and m_2. This set must, of course, be nonempty for any docking to occur at all.

Definition 63. For each entity m_1 and for each member $(m_2,[t_1,t_2])$ of Farm(m_1), m_1 is said to have committed an **impertinence** regarding the docking episode $[t_1,t_2]$ of m_1 and m_2 fif m_1 committed an intrusion regarding the docking episode $[t_1,t_2]$ of m_1 and m_2 and Involvement(m_2,(m_1,[t_1,t_2])) contains an element > 0.

Remark 60. In other words, an **impertinence** is an intrusion at an irreality level.

Axiom 48. For each ur-element z, and for each point of time t, $\sigma(z, t)$ is a positive real number.

Remark 61. $\sigma(z, t)$ is how '**bright and shiny**' z is at time t.

Axiom 49. For each entity m, and for each point of time t, $\varphi(m,t)$ is an element of Λ.

Remark 62. $\varphi(m,t)$ is what the entity m is **focused on** at time t.

Axiom 50. For each entity m, and for each point of time t, if $\varphi(m,t)$ = Deductive Mathematics, then activity_level(m,t) $\geq \omega$.

Axiom 51. For each entity m, and for each point of time t, if $\varphi(m,t)$ = Monotheistic Theology, then activity_level(m,t) $\geq \omega$.

Remark 63. It can be difficult to distinguish between Mathematics and Theology.
    cf: the following joke:
    A biologist thinks he's a biochemist.
    A biochemist thinks he's a physical chemist.
    A physical chemist thinks he's a physicist.
    A physicist thinks he's God.
    God thinks He's a mathematician.

"If a man's wit be wandering, let him study Mathematics."
-- Francis Bacon

"Lead simple lives, so that you can do complicated mathematics."
-- Russian saying

"We are not human beings having a spiritual experience. We are spiritual beings having a human experience."
-- Teilhard de Chardin

Convention 6. t = 0 refers to the present time.

Definition 64. An entity m is said to be born again fif there exists a point of time $x \leq 0$ such that for each for each point of time t such that $x \leq t \leq 0$, activity_level(m,t) defaults to ω.

Definition 65. For each entity m, and for each point of time t, m is said to have a **change of focus** at time t fif there exists a point of time $x < t$ such that for each point of time y such that $x < y < t$, $\varphi(m,y) \neq \varphi(m,t)$.

Remark 64. An impertinence typically involves a change of focus on the part of the victim. Therefore, committing an impertinence can yield an edge in a hard-fought contest. For example, Schwarzenegger relates how he used this technique against a weight-lifting rival. There is also a case I recall seeing of the high-profile tennis match that was won in this way (- rattling the opponents with some question like "Was it inbounds?" – implicitly calling their judgement into question). Both of these are refined instances of the pervasive 'trash talk' that occurs in sports.

Axiom 52. **Follies** is an infinite set.

Remark 65. Follies consists of the **items that can be posted to foreground**.

Axiom 53. For each member z of Follies, **content(z)** is a nonnegative real number.

Remark 66. content(z) is the measure of how valuable the **content** of z is.

Axiom 54. For each member z of Follies, **form(z)** is a nonnegative real number.

Remark 67. form(z) is the measure of how valuable the **form** of z is.

Axiom 55. For each entity m and for each point of time t, **Post(m,t)** is a (possibly empty) subset of Follies.

Remark 68. Post(m,t) is the set of items that entity m posts to foreground at time t.

Axiom 56. For each entity m and for each point of time t and for each member z of Post(m,t), **footprint**(z,m,t) is a (measurable) subset of the_world.

Remark 69. Posting to foreground typically requires the use / consumption of some non-personal (i.e., public) real estate. For example, placing a flag at the top of a flagpole 'publishes' the flag, but the flagpole is planted in the ground. The ground that the flagpole takes up is the 'footprint' of the flag, or, more precisely, of the publication of the flag.

Axiom 57. For each entity m and for each point of time t, **SOL(m,t)** is a real number.

Remark 70. SOL(m,t) is the **standard of living** of entity m at time t.

"We never know what someone else is going through, has been through, or is heading toward."
-- Cheryll Duffie

Remark 71. Raising your standard of living, or defending your standard of living once raised, depends on making a **quota of postings to foreground**, because you must post to foreground in order to either block intrusions (e.g., "No Trespassing") or to engage in commerce (e.g., "bookkeeping position wanted – self-starter").

Axiom 58. For each quordinal n and for each pair of distinct points p and q of X_n, and for each path f between p and q, **path_cost(p,q,f)** is a nonnegative real number.

Remark 72. path_expense(p,q,f) is the **cost** of traversing path f in the direction from p to q.

Definition 66. For each quordinal n, if M is a strongly connected subset of X_n, then M is said to be **level** fif for each p and q in M and for each path f between p and q,
path_cost(p,q,f) = path_cost(q,p,f).

Definition 67. For each quordinal n and for each p and q in X_n, p and q are said to be separated by a **no man's land** (synonymously **swamp**) fif for each path f between p and q, path_cost(p,q,f) > 1 and path_cost(q,p,f) > 1.

Convention 7. By the '**subject**' is meant the entity that we are focused on.

Convention 8. By an '**alien**' is meant any entity other than the subject.

Another topological, or at least geometric, matter relating to psychology is that of the **Bouba / Kiki effect**. (See Appendix 2.)

Also, see Appendix 3 for a note on **psychological dualism**.

Chapter 2, Section 4: Technical dissociation

With technical progress comes technical dissociation. This fact, which may be surprising, has psychological consequences, and so we need to consider it in some detail.

Here is an easy example, just to get us started. Back in the day, the top of a bottle of ketchup and its cap shared the same location, but now the bottle has a more advanced design: the cap is at the bottom, in order to enhance product availability. Ditto for bottles of shampoo. Bananas, on the other hand, underwent this dissociation long ago: the orientation in which they are handled and stored is actually the opposite of that of how they grow. Another example of such inversion attendant upon progress is the emergence of backronyms, such as SOS.

The identity element of a subgroup is always equal to the identity element of the group, but when we advance to the concept of a ring, the analogous situation does not hold.

If two entities (people, or organizations) are in communication about an event, it is usually the case that the event is of equal importance to both parties, however this is not necessarily the case. (cf: "Poor planning on your part does not constitute an emergency on my part.") Major exploitation of the unequal importance of an event to the two parties occurs in the genre of the detective story. In the detective story, one entity is the detective and the other party is the set of individuals personally following the case, which we will call the audience. The basic format is that the hero solves the case, to the amazement and applause of the audience. Everyone knows that that is what is going to happen. So the problem for the story writer becomes how to enhance this experience. Adding twists and turns to the plot is a common practice to this end. However, another technique is to have there be dissociation of importance of the crime for the two parties, that is, for the solution of the crime to be no great burden to the detective, while being a great burden to the audience. This gives rise to what we might call the 'defective detective' – 'defective' in a personal sense, not a professional sense. A crude implementation of this concept is to simply have the detective be drunk, and yet still solve the case as in Dashiell Hammett's 'The Thin Man'. This is the crime-fiction equivalent of someone being able to perform some feat 'with one hand tied behind my back'. For a lead character to appear in public drunk, but nevertheless competent, is a presentational technique that is also used in comedy. A problem with this approach, however, is that it is inevitably seen by children, who might easily be lead to believe that being drunk is 'cute' and a cool way to be. But drunkenness was, so to speak, overkill anyway, and is not needed: enter the lame detective, where we are using 'lame' as a catch-all term for any personal drawback that is not illegal / immoral such as the disheveledness of Columbo, the nervousness of Mon, and the disability of Lincoln Rhyme. In every case, the matter is far from the core of the detective, but close to the core for members of the audience.

Online, tendering payment is not a one-step procedure: you indicate your payment method, and the separately confirm that that is your choice. Similarly with changing, say, your profile information: just as in word-processing, your changes don't become effective until you separately issue the command that it be done.

Appendix 1:
Examples of application of the partition function

(By the way, r(x) is not to be confused with Ŗ.)

It sometimes happens that two connected quantities do not change at the same rate. This is true of stimulus and response, and of volume and surface area. This non-sameness of rate of change accounts for a great deal of the variability and variety encountered in the world and the universe at large. And when precisely expressed, it resolves many conflicts of focus and emphasis.

Because the science that captures popular attention is that for which the relevant parameter is vastly different from unity, and because system behavior for the extreme values of the parameter is very different from the system behavior for values of the parameter near unity, the public easily gets the impression that science deals mostly in things not relevant to practical life, no matter how much the scientists may claim to the contrary.

Example 1. Volume and surface area change at different rates. That is, the ratio S/V, where, for a given object, S is the surface area of the object and V is the volume of the object, depends on the size of the object. This one fact accounts for the nature of much of what we see in the physical world. For example, it accounts for why bugs are not just small versions of larger animals such as dogs, cats, crocodiles, and elephants, and, in particular, why biological cells are small. The Sun, being so large, has essentially zero surface area, and so is constantly spilling its guts out, which we perceive as light and other radiant energy. So, let x = S/V. Then r(x)/x is the fractional extent to which the object is mechanical, and r'(x)/x is the fractional extent to which the object is biological.

Example 2. If x is how much of a storm a given message must weather, then r(x)/x is the degree to which it is grammatically expressed. (For example: advisory signs on the roadway, newspaper headlines, and supermarket product notices – the 'storm' consisting of the fickleness of the public's attention.)

Example 3. The older a viable entity is, the more its activity consists of financing. (If x is the age of a viable system, then r'(x)/x is the fraction of its activity that consists of financing.)

Example 4. If x is the scale of the phenomena focused on, then r(x)/x is the fraction to which the description is atomistic, and r'(x)/x is the fraction to which the description is given by the laws of motion of continua. (cf: Hilbert's 6th problem: from the atomistic view to the laws of motion of continua)

Example 5. If x is the amount of mathematics that you customarily use, then r(x)/x is the probability that you consider mathematics to be a box of tools, and r'(x)/x is the probability that you consider mathematics to be an edifice.

Example 6. If x is the output level of a manufacturing concern, then, typically, r(x)/x is the probability that its returns to scale are increasing.

Example 7. If x is the number of items in an urn, then r(x)/x is the probability that there is a significant difference between drawing with replacement and drawing without replacement.

Example 8. If x is the amount of effort you have put into a document, then r(x)/x is the fraction of your effort that is devoted to content, and r'(x)/x is the fraction of your effort that is devoted to formatting (i.e., to form).

Example 9. If x is the age of a continually-updated software package, then r(x)/x is the probability that the developers of the package correctly guess the needs and desires of its end-users. (So, later on the developers fail to make the right design options for users, for example, placing the END key at the end, instead of placing the HOME key at the end. The end is a much easier place to reach, and the HOME key is more often needed than the END key, and so it should be the HOME key that is at the physical end of the row of keys.)

Example 10. If, in a game of chess, you are behind in material, and a proposed even exchange would leave you an amount x of material behind that of your opponent, then the probability that that exchange would weaken your position relative to that of your opponent is r'(x)/x.

Example 11. The partition function was known in antiquity, in the Bible story of the widow's mite. And the same mathematical phenomenon is reference by Professor Higgins in the movie 'My Fair Lady' when commenting on the salary that Eliza Doolittle is offering him. So, an offering should be judged by r(x)/r'(x), where x is the fraction of one's wealth that is being offered.

Example 12. If x is the distance of a newspaper editor from a tyrant, then r(x)/x is the probability that the editor will criticize the tyrant, and r'(x)/x is the probability that the editor will praise the tyrant.

Example 13. If x is the amount of time you spend keeping up with the news, then r(x)/x is the fraction of that effort that informs you, and r'(x)/x is the fraction of that effort that merely distracts and disorients you.

Example 14. If x is your standard of living, then r(x)/x is the fraction of your life ruled by literature (using the term loosely), and r'(x)/x is the fraction of your life in which you have liberty of thought and action.

Example 15. If x is the age of a discipline, then r(x)/x is the fraction of its activity that is speculation, and r'(x)/x is the fraction of its activity that is observation.

Example 16. If x is the distance you travel to receive content, then the amount of content you receive will be r(x), and the amount of logistics that you will have to satisfy will be r'(x). (e.g., the journey to Oz)

Example 17. If you travel a distance x away from your home, then the amount of knowledge that you get is xr(x), and the amount of information you get is xr'(x).

Example 18. If a sustainable system has size x, then its input should be divided between old and new as follows: the fraction that is new is r(x)/x, and the fraction that is old is r'(x)/x.

Example 19. homework (i.e., study in solitude) versus classroom instruction (i.e., real-time, in person): If x is the point where you are in your educational journey, then the fraction of your learning that derives from classroom study is r(x)/x, and the fraction that derives from homework is r'(x)/x.

Example 20. rise versus reach: Art consists of having the destination of your reach being a lattice point, but reach itself cannot tell the nature of a point. Telling the nature of a point requires a bird's-eye view, given by rise, but rise is bought only a the expense of reach. The fraction that you should should devote to reach is r(x)/x, where x is your amount of strength, with the remaining amount r'(x)/x being devoted to rise.

Example 21. r(x) is the amount of notes you leave, where x is the amount of notes you take; corollary: If you don't take notes, you'll leave a note.

Example 22. medium versus thing: r(x)/x is the medium-fraction of a reality at distance x from the person, and r'(x)/x is the corresponding thing-fraction. (cf: a planet as a point versus as a body, nine women unable to produce a baby in one month, Lewin's example of a cabin being approached)

Example 23. r(x)/x is the fraction of a (nontrivial) problem of size x that is inherently serial;

Example 24. If x is the age of a system, then r(x)/x is the fraction by which it is guided by live activity, and r'(x)/x is the fraction by which it is guided by literature.

Example 25. If x is the maturity of a system, then r(x)/x is the fraction of its motivation for adaptive behavior that is based on proximate causes, and r'(x)/x is the fraction of its motivation for adaptive behavior that is based on ulterior consequences. (cf: p. 225 of Heidbreder)

Example 26. If x the size of your purchase order, then r(x)/x is the probability that you will have to pay shipping costs.

Example 27. If x is the size of your purchase, then r'(x)/x is the probability that you will forget to take some (nonzero) portion of your purchase with you when you leave the store.

Example 28. If x is the amount of consideration (e.g., study) given to a given (nontrivial) system, then r(x)/x is the fraction of that consideration that involves events which are reducible to static terms, and r'(x)/x is the fraction of the consideration that involves events which are not reducible to static terms. (cf: pp. 225-226 of Heidbreder)

Example 29. If x is the degree of maturity of an organism, then r(x)/x is the fraction of its activity for which the motivating stimuli are not equal to the ulterior consequences, and r'(x)/x is the fraction of its activity for which the motivating stimuli ARE equal to the ulterior consequences. (cf: p. 229 of Heidbreder)

Example 30. If x is the size of a system, then r(x)/x is the fraction of its activity devoted to subtlety, and r'(x)/x is the fraction of its activity devoted to power.

Example 31. If x the import of an occasion, then xr'(x) is the amount of margin for the unexpected that a first-rate organizer keeps up his sleeve for the unexpected.
(cf: "A first-rate Organizer is never in a hurry. He is never late. He always keeps up his sleeve a margin for the unexpected." – Arnold Bennett)

Example 32. The utility of an amount x of money, or an amount x of technology, is r(x) (cf: how similar the first cars were to cars a hundred years later) A lot of trouble is gotten into by presuming that utility increases linearly with wealth.

Example 33. If x is the maturity level of an organism, then r(x)/x is the fraction of its environment-related activity devoted to adapting to its environment, and r'(x) is the fraction of its environment-related activity devoted to transforming its environment.

Example 34. If x is the amount of development of a system, then r(x)/x is the fraction of its development that is a matter of shifting stresses and strains – of dynamical interaction, and r'(x)/x is the fraction of its development that is a matter of rigid arrangements. (cf: p. ??? of Heidbreder)

Example 35. If x the level of your development, then r(x)/x is the fraction of your developmental activity devoted to mining, and r'(x)/x is the fraction of your developmental activity devoted to manufacturing. (cf: "The path to a better life is more of a mining operation than a manufacturing one." -- George Pransky)

Example 36. If x is the length of a text, then r(x) is the amount of information in the text. For example, in the text 'Tuesday, May 15th, 2018", the specification of the year (2018) contains much less information than if the day of the week (Tuesday) had not been specified.

Example 37. If x is your level of maturity, then the fraction of your reading-activity that consists of learning to read is r(x)/x, and the fraction of your reading-activity that consists of reading to learn is r'(x)/x.
(cf: "You 'learn to read' up to grade 2, and from then on you 'read to learn'.")

Example 38. If x is the level of development of a science, then r(x)/x is the fraction of the effort expended on speculation and preoccupation with values, ethical or otheerwise, and r'(x)/x is the fraction of effort expended on observation, paying attention to facts as facts." (cf: p. 148 of Heidreder: "the transition from mental philosophy to science, learning to rely less on speculation and more on observation, to be less preoccupied with values, ethical or otherwise, and to pay more attention to facts as facts"

Example 39.

r(x) = amount of income from an amount x of engineering;

r'(x) = amount of income from an amount x of art

Hence: in Art, you are great, or you are nobody. But in Engineering, you can make an average salary from being an average engineer. Hence the joke: "What did the art graduate say to the engineering graduate?" – "Would you like fries with that?") Also, cf: "dues-paying" that would-be actors go through (e.g., by waiting tables)

Appendix 2:
Bouba and Kiki across the divide

The Bouba / Kiki effect, being a psychological matter, and being a topological, or at least geometrical, result, can be regarded as an echo of the approach to psychology advocated by Kurt Lewin in his book 'Principles of Topological Psychology'.

Taking note of the Bouba / Kiki effect took place on the Humanities side of the cultural divide described by C. P. Snow. However it has a parallel, and potential usage, on the Sciences side of the divide, namely, in Graph Theory: Bouba corresponds to K_5 (the complete graph of five vertices), because of the relative 'roundness' of both, whereas Kiki corresponds to $K_{(3,3)}$ (the utility graph), because of the relative 'angularity' of both. So, 'Bouba' and 'Kiki' could be used, respectively, as nicknames for K_5 and $K_{(3,3)}$.

There is also shared asymptotic behavior. That is, just as the probability of an instance of either Bouba and Kiki occurring in a discourse, whose length converges to infinity, converges to unity, so also a graph randomly augmented towards infinity will, with probability converging to unity, contain either K_5 or $K_{(3,3)}$.

A piece of terminology can also be used in the other direction, namely, that discourse that contains neither Bouba nor Kiki can conveniently be said to be 'planar'.

Appendix 3:
Psychological Dualism

As is well-known, points and lines can be interchanged in geometry. This is called dualism. Each regime is called the dual of the other. Such pairs of dualities also occur in psychology. We will consider here a couple such pairs that are mediated by linguistics constructs.

Example 1. 'Frankenstein' and 'The Thin Man' are literary duals. In the former, the expert launches the career of the cadaver, and gave the cadaver his name (via authority of the public, the critics be damned), whereas in the latter the cadaver launches the career of the expert, and the cadaver gives the expert his name (via authority of the public, the critics be damned).

Example 2. Poe's 'The Raven' and Carroll's 'Jabberwocky' are literary duals. In the former, the strange creature seeks out the hero and remains forever. In the latter, the hero seeks out the strange creature and is afterwards forever on velvet. In the former a single common word is uttered by the creature. In the latter, a barrage of strange words is uttered by someone other than the creature.

In both cases the following interesting research project suggests itself: find out if the two members of a duality resonate in the same way (or, perhaps, in reciprocal ways) with people.

The term 'dualism' also refers to a dichotomy such as that between good and evil, exemplified in the story of Dr. Jekyll and Mr. Hyde. We note one further such dualism: that of 'big brother', which can mean a powerful resource in your corner, or a meddling tyrant.

(end of document)

Made in the USA
Middletown, DE
26 May 2018